OBSERVATIONS PERCEPTIONS & QUESTIONS

OPQ

by

Shoab Kamran

authorHOUSE®

AuthorHouse™
1663 Liberty Drive, Suite 200
Bloomington, IN 47403
www.authorhouse.com
Phone: 1-800-839-8640

First published by AuthorHouse 7/24/2008

ISBN: 978-1-4343-8521-5 (sc)

Printed in the United States of America
Bloomington, Indiana

This book is printed on acid-free paper.

TABLE OF CONTENTS

LETTERS TO THE EDITOR

WHY I WROTE THIS BOOK?
NEED ANSWERS AND ENLIGHTENMENT FROM EXPERTS

Observation:

Throughout my life of over 60 years, I have been observing happenings around me, perceiving the motives of other human beings, as well as observing the world events and trying to understand the motives of the religious, political, and the powerful personalities of the world. I tried to fathom it all during my life. Of course, notions of good and evil, straight and sinister, rich and poor, life and death, life and after-life, had been shaped for me by my parents, the religious/cultural background I grew up in, and then later by my living in four different countries of the world.

Perception:

Seems to me human individuals having keen observation of events, and possessing intense perception of the reasons behind these events have pondered for solutions to better human life. The creative minds used this knowledge to produce solutions for individuals and/or for the society as a whole.

In my limited, mental capacity, education, and background, I am plagued with so many questions on the contradictions between the words and the actions of so many people in the world, that I want to ask these questions in public in this book to get simple, down to earth answers for myself and for about 60% of the impressionable and not so bright (like me) population of the world. So many of the life's general demands have become and increasingly becoming so complicated for health, education, shelter, communication, politics, investment, taxation, inheritance, death etc. that I am unable to keep up with the complexity.

Seems more jobs are being created for individuals creating complexities in everything, and in turn for individuals who simplify these complexities for a price for the common folks. I do not think it is necessary to have it this way. Yes, there are complex issues, problems and solutions, but it is the duty of those who understand these complexities to describe them in simpler understandable terms for the masses. And in some cases it seems that simpler issues are twisted intentionally into complexity so that ordinary folks would have to take for granted the official versions of the matter. Of course there are many more issues confronting the general public every day. I am not a scholar, a religious leader, a sociologist, an anthropologist, a philosopher or an educator. I am just an ordinary layman and want to pen a few of my Observations, Perceptions and Questions, to see if experts would have simple and honest explanations and solutions to satiate my curiosities and for many others like me. Else, this existence would just be too OPaQue(OPQ) for me.

Question:

Are there others in the world as confused about complexity and duplicity in the world affairs and events, and are they searching for simple and honest answers like me?

TOPICS – Universal

ARE ALL HUMANS CREATED EQUAL?

Human Equality & Basic Rights, Differing Abilities / Disabilities

Observation:

All humans are not created equal but equally distributed with different physical, intellectual, sexual and emotional attributes within each ethnic group.

Perception:

Intellectual, emotional and physical attributes are distributed almost in equal proportion in all ethnic groups but not equally for all in the same ethnic group, if given the same environment for development and nurturing. That works very well mostly as each society needs people with different capabilities, initiatives and ambitions, to fill the different professional, occupational and academic requirements for the society's survival.

Human nature has a basic animal instinct to view anyone having different looks, language, religion or sexual orientation, with prejudice, and to prevent such individuals from any proximity to the property they own or to their loved ones.

In a free society with equal rights for all, especially in the USA, this freedom and equality may have been taken to the extremes. Preferential treatments are bestowed upon minority groups of individuals whose ancestors may have been wronged before. But, most minority ethnic people migrating recently from other countries are the ones using clever means to take advantage of starting businesses, seeking employment and pursuing studies in the government institutions. The real needy descendents of the wronged minority groups are left out in most instances. Also, by following this practice, some individuals of the majority groups may get prevented from the pursuit of rightful ownership, education and occupation in some instances. In fact, wrongs committed by the forefathers of one people cannot be undone by punishing the descendents of wrong doers and rewarding the off springs of the wronged.

About 1% of each ethnic group is sexually impaired or handicapped, as is 1% physically or mentally impaired or handicapped. So, the sexually handicapped people (gays, lesbians, transvestites) have no right to assert that their orientation is an alternate way of life and that they should be accepted as such with legal marriages etc. Sure, these individuals have equal rights to have long-term relationships with individuals of the same kind in the privacy of their homes, and should be accorded the same financial rights for their permanent relationships as the ones for the married couples.

All the people afflicted with abnormal physical, mental or sexual attributes are discriminated in many societies of the world, for the pursuit of education, occupation and residence of their choosing and liking. They are taunted or looked down upon in most instances. The fact is that they are an integral part of each society and have to be accorded the same rights as the majority, to pursue their desires according to their abilities as long as they do not infringe upon the rights of others around them.

As I said earlier, intellectual, emotional and physical attributes are distributed almost in equal proportion in all ethnic groups given the same environment for development and nurturing. Of course, old tribal customs, harsher climates and inbreeding over the years may have some unique racial traits but overall as a society, the types of people in each ethnic group are in the same proportion. In my opinion, the chart below describes the approximate breakdown of the general population of all ethnic groups. The chart shows seven general categories of people everywhere regardless of their racial, ethnic, social or religious backgrounds. The intellect for the good or evil deeds is equally distributed in all.

The first three categories of people are so engrossed and absorbed in their creative pursuits that benefit the society directly or indirectly that they are devoid of any personal egos, greed or motives. They are the creative minds of social, economic, scientific, medical, engineering, scholastic and political innovations. These are the special people that have been and are the vehicles of progress for the human society everywhere.

The last three categories are the ones that are harmful to the society. Of these, category 5 is the most dangerous and has been throughout history. These manipulative and hypocritical leaders have been the dictators, conquerors, colonists and opportunists that are bent on pursuing any means to satiate their greed for material resources and power. They have caused more human calamity throughout history than happened through any other natural or man-made disaster. This kind of leadership has been and can be devastating for humanity in the long run. The category 6 and 7 criminal elements of society will always be there. Society must be watchful enough to prevent them from causing harm to innocent individuals, as much as possible. The justice system has to be effective enough to keep habitual criminals out of the mainstream of society.

Category 4 is 60% of the population that is of normal average intellect, and ambitions. This is the backbone of any society. They are content with earning their living, raising families, enjoying their social life etc. They do not have the capability, incentive or desire to pursue a career to influence or command a large segment of society. Most of them do not and cannot try to perceive the complex workings of government, political institutions and world economic implications etc. Any leader especially from category 5 can arouse their anger, interest or concern by coloring the events as if affecting their and their families' safety, income and health. Scare tactics are used in many instances to sway this segment of society into action and sometimes to wage wars where millions of lives of young men and women get sacrificed. It is doubtful if any real humane objective is achieved at the end. Most wars have been carried out for greed and power. The category 4 people, being the biggest voting constituency, are the backbone of any democratic process. They do not have time, inclination or energy to do an in-depth study of all the variables of the issues that may affect the world, the country or themselves in the long run. Any charismatic political leader or a rebel rouser can sway this 60% of the population very easily for some sinister ulterior motive.

It behooves anyone seeking a public leadership role whether in a local community or all the way up to the national or international level, to have an exemplary moral and ethical character. The public at large may not realize it, but any leader who has a tendency to assert power by violence, or has immoral private life or shady personal character, would have an adverse subconscious effect on the growing younger generation. It would somehow give the impression that violence and immoral actions can be condoned as long as desired results are achieved publicly. That sets a bad example for the impressionable younger generation and as such is not conducive to develop a just and decent society in the long run.

Category	Pop. %	Characteristics	Occupation - Profession	Traits
1	5%	- Exceptionally brilliant thinkers - Acute observers - Researchers (science / medicine) - Highly imaginative and creative minds	- Originators of religious doctrine, dogma, tenet - Philosophers of new concepts and ideas - Experimenters / Scientists producing original work	- Extremely intelligent - Super leaders - Extremely conscious of societal issues - Extremely inquisitive - Extremely analytical and creative
2	5%	- Strong believers in religious, spiritual and functional reforms - Creative individuals - Inventors utilizing the scientific discoveries	- Honest religious / political / business leaders - Inventors / innovators - Professionals - Creative Writers	- Good Leaders - Highly creative - Highly motivated - Have good intentions - Socially conscious

3	10%	- Zealots / followers of categories 1 & 2	- Highly capable/honest workers/administrators	- Morally motivated - Socially conscious
4	60%	- Common simple folks with average intellectual levels and ambitions	- Farmers / Laborers - Professionals - Small businessmen - Brokers / Negotiators - Accountants / lawyers - Unemployed / retirees	- Neither good or bad intrinsically - Very impressionable - Too occupied with own work/family
5	10%	- Sharp manipulators of category 4 folks for personal ego, benefits or sinister motives - Sociopath individuals	- Sinister religious / political / business leaders / lobbyists - Lawyers/accountants - Social/Sexual exploiters	- Manipulative leaders - Shady sales persons - Complexity creators - Beneficiaries from complexity - Cheaters & Liars
6	5%	- Individuals running shady businesses - Individuals with violent religious / cultist ideas - Psychopaths	- Crooked businesses - Human traffickers - Violent gang leaders - Drug dealers - Violent fanatics	- Shady characters - Inclination/capacity to commit crime/murder - Rehabilitation possible for few
7	5%	- Inherently violent, rebellious persons - Societal misfits	- Mass Murderers - Violent Robbers - Rapists / killers	- Criminals-no remorse - Must be removed from society when caught

There is no way to infer from the above table that people in one category are destined to stay in that group only. People with strong will power and ambition have achieved miracles in spite of the circumstances. It is hoped that eventually, the first three categories of people would have such an influence on the 60% of category 4 population that the violence, dirty politics and corruption espoused by category 5 persons will not be able to influence them effectively. That way 80 to 90% of the population would strive to live in a moral and ethical manner so as to have peace, harmony and justice for all in the society. And, thus the sanity would start prevailing everywhere.

Also, the myth and old beliefs that children born of parents in any status would inherit the same abilities/disabilities and traits as their parents is completely false. It is a known fact that siblings born of the same parents can have very different levels of intellect, emotions, ambitions and moral/ethical character. Some genetic makeup certainly gets passed on to the offspring but which genes will be dominant in each child is not certain. A creative genius can be born to uneducated parents. That is why, until scientific research reaches a level to change and alter the genetic makeup of each child (and, do we want to do that?), the society must strive to provide suitable and almost similar nurturing and educational environment for all children.

There are people in certain religions and cultures and even in the sophisticated Western society, who believe that off springs of certain ethnicity, race or profession, are destined to stay within their parents' level of achievement and placement in the society. That myth needs to be destroyed. It is such a shame that the generations of people who do the cleaning of latrines and sweeping of homes and the outside, have been and still are placed in the untouchable caste in the Indian sub-continent. To me they provide almost as valuable a service for the general health of society as the doctors and nurses. And, they should be respected as much for that.

The right to realize to fruition the potential you are born with must be recognized and accepted all over the world, for all human beings. And, that is what would ultimately bring the true equality for all humans in the world.

Question:

Can we, respect all humans regardless of their ethnic, cultural or religious background, and, the people with physical, intellectual and sexual disabilities? Can we educate society in general to treat all with empathy and compassion as individuals of the same one human race?

HOW DID WE GROW TO BE DIFFERENT?
FAMILY, CLAN, TRIBE, SOCIETY, ETHNICITY, LANGUAGE

Observation:

It is amazing how humans throughout the world have and still do create affinity and closeness to their surroundings, families and neighbors, and develop unique languages, behaviors, rules for personal and business relationships, rules for personal and sexual relationships etc. Yet, their first instinct is to get rid of any stranger that looks different in attire, race, religion etc., and to treat that person as an intruder and destroyer of their way of life.

Perception:

With the world virtually shrinking fast in terms of inter-dependence on material, human and economic resources, the world communities need to find more ways that bind us together as human beings and explore methods to improve the co-existence practices for the benefit of all parties.

In my opinion, it was just a few thousand years ago that diverse individual segments of humans throughout the world grew slowly into larger numbers by learning and developing newer methods of survival through trial and error. The needs for communication, food, clothing, shelter and transportation were developed according to the resources available in each community. About ten thousand years ago, it seems, humans in different parts of the world learned enough tricks to survive the climate and harsh surroundings, and to multiply into greater numbers. This led to an increase in the development of language, agriculture, medicine, architecture, transportation, and military tools. Soon, families grew into clans and clans into tribes and ultimately tribes into bigger societies. Thus, distinct languages and cultural practices developed in each tribe. Later, powerful tribes subjugated weaker ones and as such built a larger society granting multiple social statuses and occupational levels to groups of people. At this juncture of history, exploitation by aggressive and powerful individuals and societies, for the lands, resources and people, started in most parts of the world. Scantily populated areas with harsher climates developed differently with different food, clothing, shelter and community norms. However, at the current juncture of human history almost all individuals of the world are tied into a single entity with the modern means of communication etc. It must be possible to enlighten and educate the whole world communities to live in peace and harmony without bloodshed and warfare.

Even within each of the two hundred or so existing countries of the world, different people are struggling for independence to keep their religious, cultural, spiritual, social and linguistic traditions alive. Strong suppressive measures are taken in most instances to suppress such ambitions. So much effort and a large amount of material and human resources are expended on both sides to achieve their objectives by force. In my opinion, as long as each of these independence seeking groups do not infringe upon the rights and practices of other people and countries, and agree to treat their own people with dignity, equality and respect, they must be allowed to decide for their independence by referendum in their territories. Let there be two thousand countries in the world if the people so desire. These independence-seeking groups must realize, however, that a separate country requires expensive resources both financial and human, to run the infrastructure of a separate country. One option for them may be to ask for a more independent role while still part of the mother country. Rich or poor, all countries are becoming mutually dependent on each other for sharing natural/financial resources, and for sharing human physical/intellectual labor. Rich or poor, each country would be bound economically with others for the mutual benefit of all. Of course, the powerful, richer countries may have to contribute more than the smaller less resourceful countries, but that should be the moral international attitude adopted. This way we all can keep our distinct cultures, languages and decent traditions alive, without spilling blood of youth seeking independence in so many places of the world.

Human ideological wars also have destroyed millions of lives of the warring parties in recent history. However, underlying elements in all these conflicts are mainly greed and power for the long term perceived prosperity and hegemony of the winning party. Personally, I believe, to shed the blood of one single young soldier in such pursuits is inexcusable and immoral. In many cases no victory or real outcome of the conflict is achieved and the warring parties may start good relationships anyway soon afterwards. That just means that millions of families whose young ones lost their lives, sacrificed their loved ones for nothing. Recruiting anyone younger than 55 years old to fight pre-emptive wars or undeclared wars, must be banned by an international treaty. Let older people who have already lived out their youth, fight such wars if they feel justified. The only real reason to fight war must be if some country declares war on you, or attacks your homeland. To start wars just because of some irresponsible private elements from another country cause human and material destruction in your country without declared knowledge or official sanction of that country is irresponsible and reprehensible.

Proper diplomatic measures must be pursued to apprehend and punish such individuals once proven guilty of such crimes. Of course, preemptive wars on perceived pretexts or suspicions are also just as bad. This practice can take us back in history where wars were carried out just because you are strong and have too much greed for power and find any excuse to conquer other countries.

Question:

Can humanity reach a level of enlightenment eventually, where all groups of people are obligated into negotiating the differences with each other rather than shedding the blood of each other? Can existing countries of the world, allow rights of independence to people everywhere that desire it to keep their culture, language and traditions alive? Can we still work cooperatively to help these new countries to survive and prosper if it happens?

WHAT IS NECESSARY FOR SOCIETAL SURVIVAL?
FUNCTIONAL NEEDS, SPIRITUAL NEEDS, POWER & GREED

Observation:

Looking at all organized major religions in a basic worldly manner, one has to come to the conclusion that these are nothing but rules of conducting personal, public or business life in the society to prevent chaos and anarchy. The initiators of all religions were brilliant thinkers and reformers of societies at the time. These rules derived from religions gave a code of conduct for the functioning of the society be that at a tribal, village or city level. Of course societies not part of the major religions also had functional rules for the society's existence and survival. However, most major organized religions, Christianity, Islam, Hinduism, Judaism, Sikhism etc., not only provide the functional rules but also the spiritual rituals and beliefs that bound the members of a society into uniformity and obedience to a higher divine authority. With all the doubt and uncertainty humans have about the future, safety and prosperity for themselves and their families, the belief in a divine, ever living and all-powerful entity is much more easily acceptable than taking such dictates originating directly from a mere mortal human being.

Perception:

Once families grew into clans and clans into tribes and tribes into bigger societies, the functional rules for the social, sexual, ownership, trade and other interactions needed to be set, else the random wanton impulses of humans would easily destroy the whole community through rash actions and violence. These functional rules that got translated into laws and morally just codes of conducting life in a society, are the first requirement for a society's survival. without which the society would destroy itself through chaos and anarchy.

Humans with all their innovations, achievements and abilities are still never in full control of their destinies. For most humans, inopportune and indiscriminate happenings of life and death, health and sickness, abundance and famine, achievements and failures, goodness and evil, create a feeling of vulnerability. There was continuous struggle by thinking individuals in each society to find some plausible explanations or answers for all these happenings. That is why, spiritual explanations and views have been provided by the thinkers and leaders in each society of the world in the form of religions, cults, witchcraft, faith healing, moral philosophies etc. They try to give some form of shortcut answers to the complicated questions that are beyond the scope of human physical perceptions to fathom. The beliefs in common spiritual ideas and rituals create a unifying element for the diverse communities. Thus spirituality developed as a second requisite to bind the individuals together from different segments of society.

Within the last ten thousand years, brilliant thinkers and reformers of societies have forwarded rules, values and rituals for both the functional and spiritual requirements of their society. They were rightfully accorded special respect in our societies as prophets, religious gurus, and in some extreme cases as gods. Again, the concept of greater spiritual divinity being all powerful, setting the rules of good and evil, moral or immoral acts, and being the one meting out ultimate justice for the wronged, was and is much more believable by the masses, than to elevate a temporal human being to the eternal divine status. True or not, the religion certainly gives short cut answers to the bereaved, the unfortunate and the poor of the world to be content with their misfortunes. And, it also helps them to find solace by praying and hoping for the positive future. Sixty to eighty percent of the population is readily willing to accept the circumstances as the doings of a divine being who has own purpose and agenda not understandable by the lowly humans.

Unfortunately, there were and always are sinister and exploiting clever individuals in a society that would manipulate and mold these societal rules to satiate their warped sense by twisting what is best for the society and the world when in fact it is their ego, greed and pursuit of power that prompts them to such extreme measures. They can incite the masses against other religious, ethnic or regional entities for supremacy, control and power. Many of these exploiters or extremists have created modifications to the main religions to create separate offshoots to serve their own purposes. And, therefore, there have been created so many different modified versions and cults in each major religion that you sometimes wonder if God is really in control.

It is interesting to see that cultures created by those religions that have both functional and spiritual rules and rituals, have survived much better than the ones with only functional or only spiritual rules of conduct. The religions have been used over the years by clever religious and political leaders for unifying and controlling diverse communities of different ethnicities throughout the world.

These rules whether functional or spiritual, need to be updated or modified in light of the continuous advancement in the scientific, material and human knowledge. The social and physical achievements have accelerated considerably in recent times. The functional rules of society have to be updated to deal with the circumstances and challenges of continuously evolving world societies. The sensible humans in every community realize the diversity among them, and understand that different levels of patience and efforts are necessary to stamp out archaic customs and rituals, everywhere.

Of the organized religions that had an adequate original written code for society's both functional and spiritual rules and rituals, Hinduism and Islam have to be the more elaborate ones. Even though Judaism also had both disciplines, its restriction on accepting non-Hebrew into its doctrine, limited its scope of influence. Christianity originating from the Middle East, and starting as a spiritual tenet by adopting some functional rules from Judaism, got exported to Europe by the Romans and spread quickly in different forms all over Europe, as there was a dire need of a unifying spiritual code. And, many practiced functional rules and rituals were adopted and forwarded by the European zealots of early Christianity and/or modified to fit the needs at different times.

Buddhism has been a spiritual and peace loving, non-violent religion without advancing or forwarding much on the functional rules of the society. That is why, Buddha, having his origin in India, did not have much following in India as existing Hindu religion already provided a very extensive balance of functional and spiritual rules and dogmas. So, Indochina and other far eastern cultures adopted Buddhism more readily for spiritual salvation while they already had adequate functional societal rules but lacked divine spiritual guidance.

It is disheartening to note that over time in every religion the followers have developed so many varied and intricate, religious symbols, rituals, dress codes, and customs that all of these overshadow the real message of the original simple, actual, moral and behavioral codes of each religion. There are more efforts expended in finding ways to distinguish religious adherents in looks, dress or worship from other religions than to find common decent teachings from all religions, that bind us together as one single human entity.

Of course, over the years, some rules had to be discarded in Hinduism, like the caste system or ritual of sati (surviving wife being burnt alive with husband's body). Same way in Islam, sexual infidelity by married individuals may not be punished with public stoning to death any more. In many sects of Christianity, divorce and homosexuality is accepted now. That just means that good functional rules from all religions and societies should be discussed, and eventually one whole common set of rules should be generated so that all the people in the world could strive to live in peace, harmony and a common understanding of each other.

The political processes of the Western nations may be enlightening in some respects but they have not achieved universal equality, welfare and decency for all their citizens either. Actually, human equality was not even a society's functional rule until Islam's teaching of human equality practiced by the Arab traders and conquerors spread in different countries of the world. The Moors in Spain brought the real concept of equality to Europe. Instead of forcing the Spanish population into changing religions, an assimilation of people of all faiths was accepted there. At one time Moors had a Christian priest as an ambassador to Germany and a Jewish foreign minister of Spain. After the fall of Moors the Spanish Inquisition happened and even most Spanish Jewish population was given refuge and safety in Muslim countries like Morocco and Turkey.

Therefore, it behooves all nations and societies to be cognizant of the diversity in other cultures and to be patient with those who could take longer to reach a decent common level of enlightenment. Using force or violence to quicken the pace of such change, creates a resistance and violent backlash that drags the whole issue further back.

Question:

Can humans of all persuasions everywhere reach a level of agreement to create common functional rules for all humanity to follow, and yet leave individuals to practice their spiritual rituals their own ways, as long as the rights of others are not infringed upon? Can violence and force be stopped as a means of spreading Western notion of equality, democracy and goodness to those societies not yet ready for it?

IS DEMOCRACY THE NEW UNIVERSAL RELIGION?
PERFECT FORM OF GOVERNMENT

Observation:

Just as so much human blood has been shed in the past by some organized religions, now democracy is being used as a reason to forcibly subjugate other parts of the world to accede to the dogma of democracy to appease the powerful.

Perception:

Seems to me there were no religions or civilizations about ten thousand years ago. Ten thousand years is such a small time span compared to the time of human existence on earth. As I perceive and understand the human existence, humans developed within different physical environments over the millenniums in different parts of the world. Originally being very small in numbers they pioneered into vast open spaces in search of food and shelter. I can see that couples would settle where the resources for survival were available. There is a possibility that being small in numbers and devoid of any rules for survival or societal functional rules, there was incest commonly practiced and the stronger in the group would forcibly take what he or she desired. As more enlightenment, consciousness, and learned tricks and tools for survival were accumulated, more of the humans in each group started surviving. Soon the families developed into clans and tribes, and then bigger societies of somewhat diverse groups came into existence. So a set of rules for physical functioning of the society, modes of dealings with each other, for relationships (personal, sexual and otherwise), and, for ownerships and possessions, had to be developed in each society. The society's survival would be jeopardized by anarchy and chaos otherwise. As these rules were being developed and human mind was getting more sophisticated, there was another dire need to explain or understand, life and death, random misfortunes and catastrophes, the nature and universe around us, etc. Humans in the meantime also developed languages to communicate. Where there was a bigger concentration of people clustered together the languages got developed more extensively. With consciousness increasing in all societies with the material progress, the search for spiritual well being continued. Power and greed being the basic instincts of human societies, the stronger ones would go in search of bountiful resources elsewhere and loot that from the weaker groups. Even though rudimentary societal functioning rules were set by now, the societal spiritual needs were lacking.

But, to bind together a larger group of people from different tribes, societies etc., it was the religions that gave spiritual and functional rules as if set forth by a divine entity. There have been different forms of government throughout history, but until about two hundred years ago, autocratic rulers ruled most countries. The Athenians initiated the Western idea of democracy more than two thousand years ago. However, even then only the privileged could vote while more than half the population was enslaved to do the bidding of the ruling class. In Islam the functional rules for governing were set as a body of societal elders sitting together and negotiating the issues to make decisions by consensus. Unfortunately, this was mostly ignored throughout Muslim history.

The Western idea of democracy is fine as a concept but has many flaws of its own as well. There are different types of democracies functioning in the world. The idea of 'one person one vote' sounds great as a concept, but functionally, it seems to have many drawbacks.

One, the countries that lack freedom of self-determination and education for most of their people cannot be expected to start an effective democratic process right away. The countries like Pakistan, Central Asian countries, some African and South American countries, have a large illiterate poor population subjugated and controlled by either the feudal or war lords. Until a semblance of freedom of expression, decent level of education, and economic well being is achieved by the masses, they would be forced to vote for their masters every time, and as such democracy becomes just a sham in these countries.

Two, any charismatic, clever and manipulative leader can persuade masses to vote for him or her even though personal character, private life, and expertise of that person may be lacking for such a responsible public position.

Three, once a leader is elected to an influential public post, that person could create enough clout and influence on others to get elected repeatedly regardless of the performance in the job.

Four, democracy and capitalism go hand in hand. To run an election successfully requires a fair amount of resources. Therefore, it is only the rich and powerful people who can venture into costly political campaigns especially in the poorer countries. Even in developed countries it is the big capitalistic institutions that can influence a candidate by donating directly or indirectly, large amounts of money for a candidate that would do their bidding or at least help their cause or business. These businesses also hire lobbyists for the candidates that would support their motives.

Five, democracy in the United States is a beacon of hope for the rest of the world, yet it has some outdated practices that do not make it foolproof against unfair manipulation in the party nomination and election processes.

However, as democracy is touted to be the ultimate in providing equality and personal freedom for all the citizens, then why are restrictions being imposed continually on the terms of political office holders? Does not the equality mean that whoever wants to stand for a public office should be allowed to as many times as he or she pleases? And, so it would be up to the common people to utilize their intelligence, freedom of independent thinking and voting rights to choose who ever they please. Or, do the proponents of democracy think that the common folks can be coerced easily to vote for the incumbents? If so, then definitely the theory of freedom and equality to stand for a public office and freedom to think independently to vote for an incumbent or newcomer does not hold true.

With all its faults democracy still seems to be the best form of choosing leaders who would set societal functional rules to safeguard the basic rights of all citizens. However, parliamentary, presidential or other forms of democracies need to be further scrutinized, and the rules modified so that it can have a universal appeal for all the people all over the world.

Question:

Can all forms of functional societal rules (from all religions and cultures) be replaced by one set of rules of democracy for all communities of the world, to provide equal opportunity, freedom and justice for all individuals?

Could these democratic processes allow complete freedom for individuals to practice their spiritual rules, rituals and cultural customs as long as the rights of others are not compromised?

ARE CONSPIRACIES REALITIES OR IMAGINATIONS?
SINISTER MANIPULATIONS OR JUST ACCIDENTS

Observation:

So many drastic world events occur and have occurred mostly with loss of precious and important lives of prominent personalities in the world that one tends to wonder if there is more to it behind the scenes than just the perpetrators of the crime. The real instigators and maneuvers of the warped minds, that carry out such crimes or perpetrate such actions, seem to never get implicated or punished.

Perception:

The personalities that are working to make changes in the world events or helping change the lives of large groups of people become vulnerable to be the target of violent demise especially in the volatile or violent areas of the world.

These thoughts may occur in many minds, I am sure. In most cases such events do stop the course of action being pursued or carried out by the prominent personalities that are destroyed. Where as human mind is capable of such great creativity for goodness, there are sharp, evil and sinister minds playing tricks to achieve their own motives and destroy any hindrance in their paths, by hook or crook.

The greatest conspiracy if there ever was one, was the events created that led to the crucifixion of Jesus Christ, more than 2000 years ago. The world goes on without such great personalities taken away with legacies that are left behind. There have been such sinister events happening throughout the history of mankind. Where as human mind develops and achieves creativity for goodness there always are equally sharp and evil minds trying to destroy such goodness for their own ulterior motives.

Just in my lifetime, so many events have occurred that have left grave questions in my head about the real reasons or motives. There may be clever and sinister persons who push the impressionable minds to commit such acts, or promise a reward to them financially, or may warp their mind hypnotically to act on some cue.

An assailant who was immediately killed by the police without any further investigations or inquiry gunned down the first prime minister of Pakistan, Liaquat Ali Khan. A Hindu extremist gunned down Mohandas Gandhi, leader of the famous non-violent political movements. Mrs. Indira Gandhi and her son Rajiv Gandhi, both prime ministers of India, were assassinated. Two prime ministers of Bangla Desh were killed while in office. Soon after the victory in Afghanistan by the American and Pakistan secret service backing of Taliban and Osama Bin Laden to oust Russian control, President of Pakistan, Zia Ul Haq, was killed when his airplane was destroyed in the air with American ambassador and an American General on board. No detailed investigation of the event was ever carried out. And, Benazir Bhutto, the ex prime minister of Pakistan was assassinated and hastily buried without any proper autopsy. Why her husband refused a thorough autopsy to ascertain the cause of death is beyond my understanding. Is it to use the suspicions on the cause of death for political purposes?

A supposedly deranged military recruit assassinated President Anwar Sadaat of Egypt. Prime Minister of Israel, Yitzhak Rabin, a very sincere and honest leader, was bent upon allowing the creation of a Palestinian state. He was assassinated by allegedly a deranged gunman. The nephew who had recently returned from the United States assassinated his uncle the Saudi Arabian King Faisal, who started OPEC and initiated control of oil flow to the world. Yasser Arafat, Palestinian leader, got a sudden mysterious disease and died soon after. Ariel Sharon, prime minister of Israel fell into a long coma soon after his strong directives to extradite Israeli settlers from the Palestinian territories. All these events leave certain question marks in the mind as to who or what was really behind these events.

President John F. Kennedy, and his brother Robert Kennedy vying for presidential nomination, were both assassinated by individuals who did not seem to be bright enough to plan and execute such acts just on their own. Martin Luther King and Malcolm X were assassinated without any real investigations on how, why and who directed such actions. The accidental death of Princess Diana and Dodi Fayad of England is just as mysterious and intriguing.

The plane crash and death of John F. Kennedy Jr. along with his wife makes one question about this mysterious event. One can wonder how John Kennedy Jr. with his charisma and leadership qualities, would have stood out among the current crowd of politicians, and what force he would have been if he had lived and entered politics as was rumored he would? Quote, "Just days prior to his death, NBC Dateline hinted that JFK Jr. was considering an entry into politics, and mentioned that a story to that effect was to be published in the July 26th, 1999 issue of NEWSWEEK."

So many other incidents have happened and many other important and not so important political or social personalities destroyed that one wonders if the real truth will ever come to light about the real instigators behind such incidents. These mysteries stay hidden and human minds keep questioning the real motives. Since no truth will ever come forth there is only conjecture and suspicion left to contend with.

Question:

Are we to continue accepting the facts as are presented by the powerful sources of information as they see fit to feed us not having ourselves the mechanism and resourcefulness to reach the real truth behind such sinister and apparent conspiratorial events?

IS ENGLISH THE UNIVERSAL LANGUAGE?

EASIEST TO LEARN, HARDEST TO MASTER

Observation:

English seems to be the second language of choice to learn worldwide not just because of the USA and UK domination over world commerce and affairs but because it is easiest to learn for communication in general. Yet, with its idiosyncrasies, it is hardest to master.

Perception:

As I mentioned elsewhere many different languages developed in different parts of the world for communication locally as tribal dialects and then grew into functional societal languages for commerce, domestic and international affairs. Some languages are still in fairly pure format. Many other though have been modified, amalgamated and made into functional means of communication among diverse linguistic groups. Of course, many primitive languages were only for oral communication and no real transcription form was devised for them.

With my limited knowledge of world languages and what I have experienced with my mother tongue (Punjabi), then the national language (Urdu), and later the language used for higher education and work (English), I am more convinced than ever that English language has the simplicity for quick learning than any other language in the world. I tried to learn my ancestral language (Kashmiri) and was flabbergasted with its intricate adjectives and verbal implications for each object's gender, whether living or inanimate. The symbolic languages like Chinese and Japanese are even harder and just the inflection in the expression of the same symbol can result in a different meaning many times. In Arabic the verbs not only have to be unique for feminine and masculine objects but also are different for one, two or more than two.

English on the other hand can be understood by using simple tense of the verbs. In the English speaking countries there are many immigrants who learned most of their English language by just watching English shows on television. English language has neutral verbs for both genders. English does not have gender affiliation for inanimate objects and not even for many living creatures whose gender is usually not obvious from the looks. As such it is much easier to refer to these things and animals as 'it', 'these' or 'those'. This makes it so much easier for novice in English to communicate. For many who have learned good Urdu or Hindi as the second language, make frequent mistakes in using the wrong gender for the verb and adjective for an object or animal. You have to grow up in that environment to know the gender distinction of every inanimate and animate object.

English with its simplicity and easy learning is still used as an official government and civil service functional language in many countries that got independence from British colonial rule. With the increasing USA influence in commerce and political affairs all over the world, it is becoming imperative for the countries that want to succeed and excel in international trade & commerce, to emphasize learning and mastering English as a second language.

English rightfully has the status of being the universal language and should be granted that status. People all over the world and even the immigrants in countries like the USA and UK must accept the fact that they must learn English if they adopt that country to live in. In the USA so many financial and human resources are expended in providing multiple language translations, interpretations etc. that all those resources can be used to help educate these people in functional English language. What is the use of providing nine languages at the Department of Motor Vehicles offices, when all the road names, signs and rules have to be followed by a driver in the English language anyway? All those resources, and taxpayers' money can be better spent on more worthy causes like teaching the population English and helping them become more useful and law-abiding citizens of the country.

Of course with all its absorption of characters, words etc. from other languages and cultures, English has become very adaptable. Its literary authors' works over the last few centuries makes it the most read and understood language. Yet, the intricacies, nuances and idiosyncrasies that exist in English language, make it the hardest to master for most average learners like me.

Question:

Can English be accepted as being the universal language for international affairs, commerce and education, while letting people enjoy their own languages in their homes and communities as they please?

WHAT ARE THE RULES OF MARITAL RELATIONSHIP?

SEXUAL RELATIONSHIPS, MARITAL RELATIONSHIPS, TRUTH & TRUST

Observation:

For about eighty percent of the human population, personal sexual identity and eventual relationship with the opposite sex plays a large part in their life. To satiate these natural urges and desires, every society whether deemed sophisticated or not, created rules of courtship, conjugal togetherness, fidelity and trust for the procreation and stable nurturing of the young ones. Of course, organized religions and cultures have provided elaborate and colorful processes and rules for such a union between a man and a woman. In most cases it is done in front of a large gathering of relatives and acquaintances as witnesses to this solemn occurring. Even though physical, emotional and mental compatibility has to be matched before such a union, the youthful urges and inexperience leads many to rush into such relationships. No wonder many of them end up in separation. Many other tolerate the incompatible relationship for the sake of children. And, some just keep together in appearance only pressured by the cultural or religious taboos.

Perception:

Initial and lasting love in equal proportion between the two parties of a married couple is more a myth than a universal phenomenon. There can be so many other reasons, excuses and needs to get into such relationship besides the sexual attraction and compatibility. Sexual needs and urges seem to be distributed in different individuals in different proportions. However, in almost all societies, the relationship of marriage is sanctified. Yet, so many of these marriages sooner or later turn into acrimonious or hypocritical relationships. Marriage is a contract and like every contract, business or otherwise, it has underlying obligations and actions incumbent on both parties so as not to breach that contract. Sure, there is mutual love and respect in some married couples that lasts their lifetimes and there is no attempt or consideration to breach this sacred contract.

It is interesting to note that marriage is a relationship between two individuals based purely on honesty and truth with each other. This relationship relies solely on trust and loyalty to each other. Any lie or dishonest act automatically breaks that relationship and the marriage becomes a façade even if it continues.

In marital relationship, after the first lie or infidelity, the intimacy and closeness gets strained automatically and then more lies get piled up. The fact is that the relationship cannot be mended easily after that as the mistrust created by such an act can take a long time to get reversed, if at all. This relationship is not like the blood relationship among the siblings, where they may or may not have mutual respect or affection for each other but the genetic relationship stays no matter what. Marriage is a very fragile relationship and both parties need to work at it to create a mutual atmosphere of respect and trust.

And, a notion by some persons whose spouses are unfaithful, to do the same to get even, is completely irrational and wrong. Two wrongs do not make it right. And, it certainly can influence or affect the children adversely.

It is a common notion and understanding that it is mostly men who indulge in the illicit affairs. The fact of the matter is that men alone cannot have an affair by themselves. They need a female partner as well. So, in my opinion, there are just as many women indulging in such affairs as men. The difference is men are clumsy and not careful, whereas women are very subtle and clever. So men get caught easily while women do not. Also, it is the nature of men that if a woman offers herself, a man would rarely refuse but many definitely may regret it later. Women, on the other hand, being more choosy and selective, carry out the affairs with such finesse and subtlety that their spouses are unable to catch them in most cases. Even though men may know or suspect of their wives' infidelity, their ego and self-respect prevents them from going public with the issue in most instances. Women on the other hand are very quick to proclaim or advertise their husbands' affairs etc. with or without any substantial proof. Of course, there are always promiscuous men and women. And it is tragic for their decent spouses to deal with such situations decisively in many instances especially if there are children involved.

Many spouses, who indulge in infidelity and yet hide it from their partners, can be very clever in their tactics. If the spouse complains but is not very aggressive then they would tell him/her not to indulge in self-pity. Or, they would find some item, a phone number, name etc and make it into a big accusing item to cover their own illicit affairs. They act out the psychological idiom, "The best defense is a good offence". As such it preempts the partner from accusing the spouse of any unfaithful behavior.

Again, in my opinion, truth and trust are two pre-requisites of a decent marriage and love and respect automatically flourishes where truth and trust prevails.

Question:

Can mandatory comprehensive psychological tests be developed for prospective marital couples for mental, physical, social, sexual and emotional compatibility?

Can counseling and devices for birth control and contraception be made freely available so that no children are born except in a compassionate, loving and nurturing environments only to parents who are compatible?

WHAT ARE THE RULES OF FAMILY RELATIONSHIP?
FAMILY DIFFERENCES, FINANCIAL RELATIONSHIP

Observation:

Siblings from the same parents can be born with completely different intellectual, emotional and moral characteristics. They may or may not get along with each other. However, it is assumed in almost all cultures that siblings must get along with each other or at least they always owe each other financial, material or emotional support.

Perception:

Obligations of siblings to each other are wrongfully given the same level of importance as obligations of parents to their children and of the children to their parents.

Of course a lot has been taught in almost all religions that siblings are responsible for each other and have certain obligations to each other and their parents. In fact logically, it is the parents who brought children into this world. They have a moral obligation and duty for bringing up children to the best of their abilities, and to bring them up as upright, law abiding and productive members of society. Even most animals nurture the off springs so that they can become independent and self-reliant as soon as possible. All the efforts, resources and times spent on your children, are obligations of a parent. These cannot be expected with any acknowledgement or return of any kind by the parents from their children.

I think the obligations and priorities of parents, children and siblings get confused in many societies because of local cultural and financial constraints. For the parents, the children and their welfare is the first priority, then obligation to their parents, and lastly the help or support of siblings in need. Some siblings or siblings' spouses may not understand it and as such there are so many instances of bad feelings created because of wrongfully perceived expectations.

The siblings do not owe each other anything since they are not responsible for each other's existence in this world. Some siblings, because of their love and concern, may help each other out when and where necessary. But, any such help or support provided by a sibling must be acknowledged and repaid if and when possible. The clever siblings must not use this relationship to deceptively extort gains without any intention of repayment. Even though the sibling bondage is unbreakable, there may or may not be any affection or respect for each other in some instances. Yet, even in richer sophisticated societies there is a notion that a sibling has to provide the other with support and help when asked without any obligation or promise of repayment. Of course, many siblings do help each other and have good relationships. It is up to the sibling who provided the support to accept the repayment in return or not, but any help financial or otherwise must be acknowledged and a promise of repayment made.

In many cases the disputes among siblings are because of perceived or real inequitable distribution of inheritance of parents' estate and/or belongings. The grudges and disputes keep growing with time passing as in many cases maturity and sensibility to deal with them are not practiced.

It is natural for younger children to expect their parents to know everything and to make the right decision every time. In fact the parents, in their twenties and thirties, too young and immature themselves, can be struggling at times through feelings of anger, depression or inadequacy, to provide the best nurturing environment for the family. Yet, in most cases the parents do love all their children equally, and special treatment is given to one child or another only because in their opinion that child needs it.

The notion especially in the Western world that children reaching age 18 must be responsible for their own living, or pay for it if living with their parents, is not correct, in my opinion. As mentioned earlier, different children grow and mature, mentally and emotionally, at different speeds and times. Parents have to do the utmost for making them capable of living independently. Parents must not push the children out of the home prematurely unless the children are ready to face the world on their own. Otherwise, there may be a chance that the criminal elements of society could exploit them.

Just as parents' love, affection, time and resources expended on children are not expected with any acknowledgement or return, same way the children's affectionate support and help to the parents is an obligation and does not have to be acknowledged or any return expected. Also, all siblings may not provide the same affection or similar support to their parents. But, there is no counting in this relationship. Each child must do what he or she can do or deems necessary to do for the parents without any regard of what and how much the other sibling is contributing to the effort. This situation becomes more acute as parents live much longer and at later stages of their lives need much more support, care and patience from their children.

Life is so short and time marches on. All the hurtful remarks made or the retaliatory actions taken in the heat of the arguments with the siblings or the parents become so trivial in retrospective when you reach advance age. Many siblings keep real or perceived injustices and hurts gnawing at them all their life. To many the notion of forgiving, forgetting and regrets may come too late especially when the parents or siblings are mentally or physically too feeble, or gone forever from this world.

Question:

Can the parents and children cherish the good times spent together as a family rather than dwell all your life on the perceived inequities and hurts suffered at a young age, and keep hurting each other more through words or actions well into the late life?

WHY DIRECT YOUTH TO CREATIVE AND HEALTHY PURSUITS?

Rebels without a cause - Education and opportunities for all

Observation:

So much energy of young population everywhere gets wasted or misdirected towards destructive, foolish and violent actions because of the vigorous and uncontrolled harmonic, sexual and egocentric urges starting at the adolescent age. Coincidentally, those are also the most productive years for possible self- development and overall accomplishments.

Perception:

For young individuals, an older person belongs to another world, and as such unable to understand their concerns and problems. Physical decline and aging seems to move faster than we realize. The older individual is just as confused as the younger ones about many issues, but has learned to be patient and not be too impulsive. Need is to enforce the young people everywhere up to the age of 21, to be occupied with continuous creative and decent intellectual, physical and emotional pursuits so as to prevent them from wandering into self destructive territory.

Religions do direct some of the younger population away from the self-ingratiating impulses, but many get too brainwashed with religious zeal that over shadows the real human values and inculcates prejudices against other cultures and religion. The need is to produce so many intellectual, social, physical, local and international pursuits on a friendly competitive level that not much idle time is left for young people to get into any mischief. Of course, no solution is perfect but we must strive for the better future for all continuously.

Question:

Can we use creative methods to have all our youth attracted to and concentrated on the decent physical and intellectual pursuits so as to redirect their energies less towards satiating their emotional and sexual urges during the most productive years of their lives?

ARE HUMANS STILL EVOLVING?
JUSTICE AND RESPECT FOR ALL COUNTRIES AND INDIVIDUALS

Observation:

We have come a long way from even a hundred years ago when colonization by powerful nations was rampant and the basic precept was exploitation of human and land resources by hook or by crook of the conquered territories. Might, had always been right until lately.

Perception:

Because of greater need for international human and material resources worldwide more interactions and cooperation is being sought and achieved by making us move away from the short sightednes and differences over trivialities of race, ethnic and religious affiliations. There is developing a general code of behavior for all humans on earth. The world is shrinking because of the technical advances in communication and otherwise. A few hundred years is nothing in the history of mankind whose consciousness reached a level of sophistication only about five to eight thousand years ago to create societal rules, modes of conduct, religious rules, and the pursuit of power and greed. And, this led to incursions and exploitations of powerful societies over the weaker ones. Just a hundred years ago European nations had wars, conflicts and atrocities commited on the weaker members on a mass scale. Now, any such act is incomprehensible there. And, to set up a World Court to try despotic and cruel rulers is a step forward in enforcing a humane behavior for all humans. My perception is that all this is happening in the evolutionary landscape. Even though sinister forces in pursuit of power, greed and lust, are still at work, there is more of an outcry from within these powerful societies, to have an equitable, just and cooperative attitude towards less enlightened and poorer societies and people.

Not only in international arena, but also internally within the sophisticated enlightened societies, there is more emphasis being given to bring the levels of humanity to a decent living standard for all. There is more support and assistance being provided by the societies to individuals not fully capable of taking care of themselves or their affairs. Greatness of a society will be measured in history not by how powerful a country is economically and militarily, but by how well it treated its unfortunate population which is financially, mentally or physically deprived, and how well it provided the needed assistance for their everyday living.

Question:

So, are we headed to a human evolution on a mass scale for the equality and benefits of all humans living peacefully and cooperatively with each other on earth or are we headed towards annihilation and destruction of this beautiful world through violence, killing and pre-emptive strikes in the name of God and goodness?

TOPICS – American

IS US FOREIGN POLICY BEST FOR AMERICANS?
THE ONLY SUPER POWER, YET THE DESIRE FOR MORE CONTINUES

Observation:

In just the last fifty years the United States of America has gone from the most admired to the most despised nation in the world because of the way US foreign policy has been conducted. This country, having the best precepts under which it was created, has the potential to regain and sustain the international prestige that it has lost.

Perception:

The United States of America is a nation of immigrants. Where as in Canada or the countries south of the US, most European immigrants either co-habited with the natives or took most of the land through less violent means, the immigrants that came to the US territories were more aggressive, ambitious and enterprising.

These pioneers are restless souls in search of better opportunities and resources. However, in spite of the subjugation of the native population and the exploitation of the natural resources, and introducing slave trade into these territories, the thinking decent elders of the country created the best constitution and form of government anywhere in the world.

Because of the entrepreneurship, aggressive and ambitious basic nature of its continuing newer citizenship, the country has progressed economically, culturally and militarily much faster than any other nation. Strangely enough, even though the immense loss of human life and infrastructure that these wars bring about, the push to win the wars gives extra kick to innovate newer products and to find ways to improve the war methodologies. A perception of a continuous external threat whether ideological or military, real or concocted, provides two essential benefits to the country. One, it unites the country against the external threat and removes the focus from internal political or social issues. Two, it allows allocating of humongous resources to research projects to keep American military and economic superiority in the world. The wars, ideological or military, keep pushing the American ingenuity, entrepreneurship and innovations to a greater level. Many of the outcomes from such research and inventiveness become common everyday tools for average American, and gain worldwide usage, as well. Spy satellites and Internet are just two examples of transformation of innovations from military to worldwide civilian usage.

Throughout history, all those great civilizations that became complacent about their invulnerability quickly imploded and had external aggressive forces take them over. The need is to find ingenious ways to keep the momentum for competitive inventiveness going without creating wars or warlike atmosphere within the country or without. Sanctity of human life must be observed and young human lives, military or civilian, must not be sacrificed unless absolutely necessary to safeguard your home or homeland from an imminent threat.

In order to keep her hegemony, the United States had to be very vigilant so that the ideals of capitalistic and free enterprise system espoused by the US are not jeopardized by other hostile countries. The US is the only Super Power in the world right now. To achieve an economic hegemony also referred to as the United States' worldwide interests and way of life, the US Military, Foreign Service and CIA work hand in hand. Idealistic wars especially of communism versus capitalism, made pervasive foregoing or ignoring of all ethical rules by both parties to keep their dominance on other countries. Even, with the fall of the Communism in the world, certain actions are still taken by the CIA to install dictators and support rogue rulers all over the world to safeguard the US hegemony and interests. Many of these monsters created by the CIA and supported by the US Foreign Service then get out of hand, and later they have to be destroyed by the same forces that created them. Democratic ideologies and human rights are not of much concern for such foreign rulers who are doing the bidding of the US for keeping secure their power over their people anyway possible.

Another aspect of CIA's covert activities in fighting ideological wars against other countries is the hiring of local populace whose reward both on the success or failure of the project is a promise of easy entry into the US as refugees. They are provided certain remuneration for every adult individual and also government paid housing for years to come. These people who were not faithful to their own country, how can they be expected to be faithful to this country, and be a contributing and useful member of the society here? Their only purpose in life is to squeeze as much as they can from the government or other resources for their personal benefits. Most do not pay taxes as they work for cash mostly, else it would cause reduction in the government remuneration and benefits. Most do not even vote. These so called refugees are in much abundance, brought here from Cuba, Vietnam, Russia, China, Korea, Afghanistan, Iraq, Iran, Kurdistan etc. I strongly believe that such people should be paid off in their own countries for the services rendered rather than importing such elements into this country. This country needs to have strict immigration rules to allow only productive decent law abiding individuals for the safe, secure and prosperous future of the United States of America.

The world has evolved now to a state that all countries should be treated with equal set of rules anyway. Of course, there will be countries that are not friendly. But, with the world economy and interests so intertwined, it behooves to negotiate with all countries as equal. The notion of decent and equal treatment and willingness to communicate and negotiate issues would bring back the prestige that the US deserves for its inherent ideology of decency and goodness for all humans on this earth.

Constant vigilance and shrewd observation of all nations and their regimes is always needed to safeguard the country and its citizens from any evil nation or group. But, too much indulgence in pre-emptive strikes or covert activities, risks judgmental errors and loss of innocent lives and creates severe resentment and mistrust among the general population of adversarial as well as friendly countries.

The borders of this country must be made secure from infiltration of any illegal entry or activity. Of course, the country cannot be transformed into a police state else it would lose all its momentum for human progress and innovation. But, you cannot avert your eyes from illegality and hiring of non documented individuals. Those whose first act of entering into the country illegally is to break the law of the country, how can they be expected to abide by any other law of the country?

Those poor people trying to sneak into the country for economic reasons and being hired for minimal wages by the US farming or manufacturing concerns must be provided temporary legal working status before entering the country. Of course legal residents demand higher wages but moral, ethical and legal codes must supersede the greed for cheap labor.

There seems to be a subtle opinion of extreme political right wing of the country that charity receivers whether individuals or nations do not have any rights. These charity receivers cannot question or object to how the charity giver acts or dispenses charity. I perceive that to be wrong. Much more compassion and understanding of the unfortunate people and nations of the world is required. Charity must always be dispensed in a way that it is felt to be received as a payment for a legitimate service rendered. Else, the first thing that happens to a charity receiver is to lose self-respect. And, once self-respect is gone then ethics, rules and laws do not have much meaning and all becomes fair game.

Question:

As we continually pursue to respect the personal freedom and equality for all our own citizens in this country, can the US foreign policy be conducted to show the equality and respect for all countries and their citizens whether they are friendly or critical of US foreign policy, and as such eventually regain the international respect it really deserves?

WHY HAVE TERM LIMITS FOR ELECTED OFFICIALS?

ELECTORATE INTELLIGENT ENOUGH TO SELECT THE RIGHT CANDIDATE

Observation:

There seems to be a growing tendency on shortening the number of terms in office of the elected officials at the local, state and federal level. Seems, the election processes have become much more complicated and subject to sinister manipulation by the incumbents and/or the powerful elements of the society.

Perception:

What I have been made to understand is that democracy is the perfect form of government that provides equal opportunities to all the citizens to stand for a public office and allows one to vote for anyone that is deemed suitable and best.

Also, it is assumed under this hypothesis that all individuals are intelligent enough to reject the incompetent and corrupt incumbents to be reelected. Isn't the maxim, people get the leaders they deserve, true?

In fact democracy's biggest flaw is that it is very much a game that mostly rich and influential people and entities can play. Of course, persons with charisma and leadership qualities can influence enough people and accumulate enough funds to run the election campaigns effectively. Yet, the money or the promise of more, can be a great motive in public's mind to change their votes. Also, influential community leaders in cahoots with corrupt or sinister politician can easily influence the majority of the population.

It is, therefore, extremely important for the honest and decent religious, academic and social leaders to continuously keep the public aware of the kind of public service and contributions provided to the community by the elected leaders. That way, the need to have a cap on the term limits of elected official would become redundant and we could revert to the true democratic ideal of providing all the freedom to stand for a public office without restriction and let the electorate do the job of choosing the best candidate for the position.

In addition, the United States of America has made the party nominating process and the presidential election process extremely complicated by having the delegates, super delegates and other options and restrictions in each state of the Union. More fractious and complicated the process becomes, there is more chance of manipulating the processes by the influential, clever and sinister overseers of the political or electoral processes. Allocating party delegates who themselves may not be elected by popular vote, and then having these delegates use their preferences contrary to the will of the popular votes has happened at times. In my humble perception this kind of election process is outdated and flawed subject to manipulation by many ways. It may have been a good idea up to a hundred or so years ago when the universal electronic voting systems and good and easy transportation for all was not available. I think, the best, simplest and most economical and efficient method would be to use direct popular vote for party nomination as well as for the presidential election process.

Question:

Are the term limits of elected officials a reflection of a true democratic process?

Is the electorate not intelligent and perceptive enough to reject non-productive or corrupt incumbents anyway and elect good leaders through direct popular vote?

WHY HAVE TAXATION SYSTEM SO COMPLICATED?
BENEFITS BUSINESSES, LAWYERS AND ACCOUNTANTS ONLY

Observation:

The United States of America being the largest capitalist country in the world gives all the breaks to the large businesses in the taxation system so that they keep growing bigger while the middle working class pays the most taxes and gets the least benefit out of this system.

Perception:

I do not think a society can function without a local and/or central government. There are collective local issues as well as collective issues of the country. To handle these issues requires personnel and resources. There are so many collective issues and needs of a society like health, transportation (roads, air links, ports etc.), safety and others, that to fulfill those needs of a society, the governing systems have to allocate optimum resources for each of the needed functions to be carried out effectively. To that effect, different types of taxation systems are setup in every country. In European countries and Canada the taxes are high but the benefits for the common people are just as good. Cheap or free higher education, universal health care, retired senior population care, etc. are provided by the state in those countries.

However, here in the United States, every aspect of our everyday life has to be controlled and dispensed by big businesses. To that effect, all businesses and rich entities especially the big ones are provided all the incentives to use all kinds of loopholes to pay the least amount of taxes. Shrewd lawyers and accountants are employed by this privileged class to pay the least amount of taxes in order to save money to grow the businesses more. The notion or fallacy is that bigger the big businesses get, more the people get hired and as such the whole population benefits from it. President Reagan referred to this practice as the "Trickle Down Economy", which in fact is "Trickle Down Greed". This trickling works downwards only up to the businesses that are customers, suppliers or vendors for big businesses. The common workers keep on getting burdened with the complicated taxation system where their disposable income barely can keep up with the inflation and expenses.

Because the sharp taxation lawyers and accountants who draft up new taxation codes for the benefit of the wealthy, have made the whole system and process so complicated and time consuming that the only ones to get the benefits out of it are the wealthy businesses and their lawyers and accountants. And, the most job creation that happens is in the Internal Revenue Service for handling ever increasing complexity of the taxation system. Any honest politician or person who proposes flat tax to simplify the taxation system is quickly shot down. The clever and resourceful parties whose current and future prosperity is closely tied to the complexity of the taxation system would always try their best to prevent flat tax proposals from reaching any level of consideration.

Also, in addition to the income tax there are taxes for every aspect of your life. There are sales taxes, property taxes, inheritance taxes etc. On top of that local governments have fees, bonds and measures for education, traveling on the road, owning a vehicle, crossing a bridge, smoking a cigarette etc. Of course, where necessary, appropriate fees should be charged to cover the costs and emergency expenses. But, government fees, bonds and other special charges are inappropriate and confusing to the public. There is a strong need to run the governments like a family runs a household. From one income, the amounts are allocated for education, health, shelter, clothing, food, commuting and communication etc. On top of that most families put aside certain amounts for emergency or unforeseen circumstances. Why can the governments not plan and run the budgets the same way by allocating appropriate amounts for regular expenses and setting aside some for unforeseen disasters or happenings. Why do they have to issue a new bond etc. for any new or unusual expense? And, how long can we put up with the ineptness or is it the special interests of political leaders in our governments to run the allocated budgets properly?

Mainly there are two types of flat tax proposals offered. One is abolishing all taxes except the sales tax that would then be distributed proportionately to local, state and federal coffers. The other proposal is revenue based. Every business, enterprise, and individual would pay the same percentage of tax on the gross income. I think, every person or business paying same amount of tax rate on the gross income or revenue, has an advantage over sales tax. With the interstate and international shopping increasing over the Internet, it may be hard to manage the taxation receipts from such transactions. With the revenue based taxation system, businesses and investors would not get away without paying any taxes. Of course, there will always be some illicit cash only underground businesses evade paying any taxes. The revenue based or gross income based taxation system will help the middle class that currently carries the biggest burden of taxes. Same percentage of tax applied on the gross income of all businesses and persons at the highest to the lowest level will provide four benefits to the society.

One, it will generate enough tax revenue esp. from the businesses and investors that overall tax burden would get reduced for all especially the working class. Currently, this highest income group with the help of shrewd lawyers and accountants pays the least by using all the loopholes provided by the current system to their advantage.

Two, it would simplify the tax filing and management processes for all and as such IRS personnel, tax consultants, lawyers and accountants could do something more productive and useful for the society than harassing the common tax payers.

Three, it would create a real equality for all. A poor farm laborer will be paying the same tax percentage on his income as Bill Gates would.

Four, rich businessmen and industrialists would not be penalized with higher tax rates just because they were more enterprising, astute and innovative to build and grow thriving businesses.

Let equality prevail.

Question:

Can we impose on all our politicians to seriously consider a flat tax proposal in order to simplify the taxation code and disband IRS, and to prevail upon the legislature to pass such a resolution?

IS DAYLIGHT SAVING TIME STILL NECESSARY?

The industrial age over, Flexible work hours, Work from home

Observation:

The practice of Daylight Saving Time is redundant and unnecessary in this Electronic Age. Industrial Age is over and with flex work times, work from home options and staggered working hours make this cycle of artificially changing time twice a year, very cumbersome, confusing and counter-productive.

Perception:

In the Agriculture Era and later in the Industrial Age the idea of changing the clocks to one hour early in the spring and changing it back by one hour in late fall was understandable and may have saved energy resources and given more time to the families to enjoy outdoors in the daylight hours.

Now, it is the Electronic Age where majority of the young population would rather spend free time with their computers or electronic games. Of course, many would still participate in sports etc. but just to sit around in a park or open space because of the extra daylight time is increasingly diminishing. The parents also come home after work and turn their computers on to catch up on their emails and world events. As the major portion of the population spends more of their free time indoors, the saving on electric consumption will not diminish in most cases as increasingly more homes are getting equipped with central air-conditioning, personal computers, computer games and fancy audio/video arrangements.

In the late 1980's and early 90's the maximum speed limit on all freeways of the country was reduced to 55 miles per hour citing the savings in the consumption of gasoline. However, no authentic and practical statistics were produced to prove the actual savings attained. In fact, with the slower traffic during rush hours, the time spent on the roads would increase and as such the consumption of gasoline. At that time the price for a barrel of oil had gone above $25 and as such it was thought to create a crisis and the government agencies in charge had to show some remedial measures devised and incorporated. Now a barrel of oil costs over $125 but there is no talk of reducing the maximum speed limit and the big oil companies are reaping record profits. That means that 55 miles per hour max speed was just a political ploy at the time.

Similarly, non-verified arguments and figures are being promoted to continue the DST and in fact to increase DST period by almost two months starting from 2007. I have not seen any discernible decrease in my energy bill with these changes. In fact, the rates have gone up almost five times from just twenty years ago. Government agencies in charge of regulating the flow, consumption and prices of the gasoline, electric and gas commodities have to show some definitive measures being taken to control the situation. Of course, the large oil companies, utility and energy-producing corporations have too much clout and power to be controlled and regulated. So, a façade is created that effective measures are being taken to amend or control the situation. Common ordinary folks become mostly the scapegoat to suffer through these unwise regulations.

So many corporations and government departments have mainframe computers, and other types of data and computing servers and devices in operation twenty four hours of the day. Many of these especially the mainframes have to log every activity and archive it so the statistical analysis or effective problem detection can be carried out. These expensive systems' clocks have to be changed twice a year because of the DST. In most cases, changing time during spring can be done online without any stoppage. However, for the fall back of time these millions of dollars of equipment has to be brought to a complete stop for one hour so that there is no overlapping of the log data. Just the one hour wastage of these machines and one hour of non productive times of the workers would be enough to offset all the presumed savings received through this artificial playing with the time clocks.

Of course, most households have a wristwatch and alarm clock for each member of the family. Also there are clocks and timepieces in the house. Then there are clocks in every electric and microwave oven, TV accessories, cars etc. To change time on all these is time consuming. The population of this country is aging rapidly and these complicated procedures do become more tedious and seem unnecessary. That is why many people leave many of their timing devices unchanged.

In many parts of the world, some institutions like schools and government offices adjust the opening and closing times for different seasons of the year. I think, that practice is much more preferable than to change the clocks twice a year. With some places changing the business hours during summer time would also stagger the rush hour traffic and save on the gas guzzling during rush hour by not being stuck in the traffic.

Also, the government and public leaders have to move away from wanton greed and change the attitudes to teach to live in harmony with nature, and strive for economical utilization of natural resources that nature provides so as to give it time to replenish some of it. Fuel savings by cumbersome and outdated concept of Daylight Saving Time, can be realized many times over by the authorities giving incentives for producing and utilizing the alternate resources of energy like Solar, wind and bio-fuel. In my opinion, the abundance of sunshine in the poorest parts of the world, and sun's passive energy makes it the best and economic source for energy for the world. Efficient solar converters for home, automobile and commercial needs must be developed. Incentives must be provided to consumers to convert to alternate cleaner forms of energy usage.

Question:

Can we convince authorities to stop enforcing the Daylight Saving Time and join other enlightened communities of the country that have rejected such practice?

Can we recommend changing business hours during summer time for those institutions that may find it necessary to do so?

Can we impress government to provide incentives to the manufacturers and consumers for utilizing alternate forms of energy?

WHY IS UNIVERSAL HEALTHCARE NECESSARY?

INSURANCE COMPANIES - SOCIALIST CONCEPT BUT CAPITALIST TOOL

Observation:

Healthcare in this country is a mess. While the rich can afford the best care and treatment, and the poor and unemployed of the country can get good care and treatment at the county, city and university hospitals without any charge, the common working members of the society have to pay ever increasing premiums and deductibles such that many shy away completely from insuring their families and themselves for the medical and dental care.

Perception:

Insurance business whether it is for medical, automobile, property or any other coverage is in fact a socialist concept, where many pay for the few who may incur the sickness or damage. Insurance companies deal purely in money coming in and money paying out for the claims received. They do not manufacture any useful product or any other redeeming benefits to the society. All the publicly held insurance companies have to show a fair growth in revenues and profits to compete well and be viable. Once a saturation point on the number of insured is reached, the only way insurance company can show growth is by increasing the premiums and reducing the payout. And, that is what is happening with most types of insurance companies. The situation in the medical insurance business is especially acute for three reasons.

One, the clinics, the hospitals, the pharmaceutical companies and other health related businesses have to prove viability by continuing increases in prices charged for the products and services. So, the insurance companies in turn have to compensate by increasing the premiums.

Two, the growing older population of the country increases the frequency of visits to the doctors or hospitals and as such increase in the treatment of these patients demands higher premiums to cover the expenses.

Three, the pharmaceutical manufacturing companies in the US are so entrenched in the capitalistic ideals that they stand next to the big oil companies of this country in their reach to influence the political and legislative processes of this country to squeeze the maximum benefit out of their products at the expense of the people of this country especially the seniors.

It is evident that as you grow older, you want less to be burdened by too many bills, payments and budgetary intricacies. At the age of 70 or older, a person has done enough to work productively their adult life, serve the institution of their choosing, pay taxes and social security etc. Especially they, and even the younger population would like simpler medical and pharmaceutical insurance coverage options, so that you could go to any nearest doctor, hospital and pharmacy to get the necessary medical care without much fuss. That is why the HMOs like Kaiser Permanente are so popular with the seniors as they can get medications, specialists care and health guidance within the same facility. In general, for the Medicare patients the new rules, guidance and coverage options have been made so much more complicated that it benefits the medical insurance providers and pharmaceutical manufacturers only and not the patients or the institutions providing the treatment.

What generally is not realized is that the insured walk the same streets, frequent the same public places as the uninsured, and they breathe the same air. So, if the uninsured are not treated quickly and effectively for some infectious and dangerous ailment, then the infection can very quickly infect millions who are insured as well as uninsured. Therefore, the physical wellness and health of a society does not depend on only the medically insured individuals but on all individuals being treated as soon as possible for any ailment.

Unlike other types of insurance, the claims against medical insurance are more frequent and pervasive. However, if all the payouts through government sponsored coverage like Medicare, MediCal etc. are lumped together into single payer at the federal or state level, then the overall medical expenses can be reduced for all the people by reducing substantially the administrative and bureaucratic procedures and expenses. This way all can be medically covered at a much lesser expense. The burden for such coverage would not need to be increased that much higher than now for all taxpayers. Definitely, the health insurance companies must be non-profit organizations so that the emphasis is not on showing the most income (profit), but on serving the population to contribute to a healthier and productive society. If not run by a government agency, then an organization like Kaiser Permanente could be entrusted with administering this essential service with proper government checks and balances.

Single payer system would make it so much simpler and easier especially for the growing senior population of the country to go to the nearest doctor, specialist or facility for any treatment without worrying about all the formalities that are required now for the provider to make sure where the money is going to come from for the services rendered. Of course, for specialized non-critical complex exams and treatments, the patients would have to wait longer than they do now. That would be acceptable for most of the population as it is in Canada and much of Europe. Anyway, the rich who can afford it would always be able to go for instant treatment at whatever price and wherever it is available. This single payer system managed at the state or federal level would make it much more efficient and cost effective by not having many bureaucratic and administrative overhead expenses.

Question:

Can we convince our political leaders to legislate universal health care system through single payer source at the state or federal level?

WHAT IS THE IMPACT OF AGING POPULATION?

OLD TAKING CARE OF THE OLD, WORLD WITHOUT BORDERS

Observation:

Increasing education, prosperity and health care brings about the longevity in age of the population in every part of the world. More educated and self-reliant the female population gets, less inclined it becomes to breed too many children. The rich developed countries of the world already face the dilemma of having ever shrinking young population, and to cope with ever increasing retiree aged population.

Perception:

With the increase in life expectancy especially in the well off societies of the world, the whole paradigm of life cycle is changing. Just a few decades ago the life expectancy was not that high, and the working individuals of families would see their children grow up and get settled, and then would plan to retire early to have a life of leisure, travel and fun for themselves. But, all that is changing now for most of the working class families. Many of the eighty to ninety years old parents of sixty and over retiring couples are still alive and need continuous support and care. I call this new phenomenon, "Old Taking Care Of The Old". While the children grow up to be independent and self-reliant, their grand parents start deteriorating gradually and become more dependent for their well-being on their children or the society as the time passes. Many retirees now spend most of their spare time looking after their old parents rather than enjoying their retirement years.

The birth rate in the developed countries decreases with the increase in education of women and their opting to have careers outside of the home. That leaves a gap in the workforce with lack of enough labor force to cover the skilled and unskilled job demands. That is why the immigration authorities and governing bodies are not enforcing the immigration laws very strictly. This demand will keep on increasing, as the ratio of old population to the young keeps getting bigger. The unskilled labor has been tolerated in the developed countries for the last few decades. Now, the shortage in essential high skilled job market for doctors, nurses and info tech specialists etc. is creating a new wave of immigration into the developed countries of the world. Eventually, the world would have to become one entity of humans without strict country borders. The skilled and unskilled work force will move around the globe wherever there is demand or suitable opportunity.

The fundamental economic pressures to make a living by the individuals in the over populated areas of the world and the need to maintain the flow of affordable products and services in richer countries, is creating a fast paced migration of people and skills across the globe. The aging population in the richer countries would live longer and need more health and living support. That is why there is such acute shortage of nurses and health care workers in the developed countries and skilled health workers are brought into these countries in large numbers. And, because of shrinking population growth in the developed countries, there is continuing shortage of skilled and unskilled labor developing to fulfill the vacancies created by the retiring population. In spite of the growing demand for stricter immigration controls, the sheer demand and need for foreign labor is continuing the porosity of the international borders between the richer and poorer countries. Seems, eventually, the world will balance out all countries by economics, population, age etc. That would eventually create a serene and peaceful world without borders.

Newer ways have to be found to cope with the increasing aged population. Of course, the sharper business minds have started to target the well-off old population with long-term care insurances, reverse mortgages, expensive nursing facilities and senior centers etc. As far as I can see following areas need to be looked into in coping effectively with the ever-increasing population of old people.

One, there must be encouragement and availability of work for the over 60 crowd so that they stay physically and mentally healthy longer rather than deteriorate through inactivity. It may not be a bad idea to change the trend and have persons who are fifty-five or over to serve in the armed services instead of the very young. It would serve two purposes. It would provide younger population a chance to live through a productive and normal adult life rather than having their life end in its prime or suffer from posttraumatic syndrome for the rest of their lives. Also, it would reduce the burden on the social security and medical expenses and help this older population to stay physically and mentally healthy and alert much longer. With proper conditioning and exercises this older population can easily fulfill the rigorous demands of the armed services.

Two, there should be old age cooperatives started in each community that would create a self managed and mostly self sufficient facility having a self help atmosphere for the members. Since everyone gets old including doctors, nurses, business managers, teachers, gardeners, cooks etc., a barter system arrangement could be set where most common services in these senior centers will be provided by the residents themselves with minimum outside help or resources. This way the facility can be run economically. The mentally and physically healthy seniors will prolong their well-being by being productive and useful to the community. Of course, for very specialized needs, outside specialists and experts may be needed. This kind of arrangement should make such living possible and affordable for most of the senior population and take the pressure off of their children to worry all the time about their health, wellness and social life.

Three, Medicare services must be made simple to understand and use without any confusing co-payment and deductible options for treatments and medications. The seniors should be able to go to the nearest doctor or hospital and get the required treatment without delay. The simplicity, universality and availability of medical help and treatment for the whole population especially for the seniors, is urgently needed.

Four, the long-term care for the seniors is becoming a big business in this country. The insurance companies charge high premiums for it and promise so much but provide so little in many cases. A person can save money for old age but if continuous care is needed in a nursing center then all that money can evaporate very quickly. A uniform, decent treatment of such patients is needed that should be legislated by the government rather than left for the private sector to exploit the seniors off of their lifelong savings.

Five, we must reconsider and modify if needed the idea of Dr. Jack Kevorkian, to provide the aged population an option to have instructions drawn up for the end of life in case of painful incurable disease, mental dementia and/or other debilitating physical ailment. That would help the aged population to die with peace and dignity rather than go through long term suffering for themselves and the caring loved ones.

Question:

Can we transform our attitudes towards the older population in ways to help them live a productive and happy life as long as possible?

Can we let the old and infirm to decide to end their life in a dignified and peaceful manner in case of an incurable mental and/or physical ailment, so as not to be a prolonged burden on the society or their loved ones?

LETTERS TO THE EDITOR

WEDNESDAY
September 3, 2003

TUESDAY
January 13, 2004

SUNDAY
February 26, 2006

LETTERS TO THE EDITOR LETTERS TO THE EDITO

No to Issa-ism

I AM getting increasingly disillusioned with the American politics.

The recall effort against Gov. Davis, so soon after his re-election, is bizarre and absurd. He has impeccable qualifications and experience to administer the state's affairs.

Unfortunately, he is not a clever politician and lacks charisma to excite the media or the public.

Compared with many other Democratic or Republican politicians working directly or indirectly for special interest groups or their own interests, Davis quietly has been running the affairs of the state in the best way he could — considering the dire economic downturn and energy constraints created by the federal government's misadventures.

Seeing Arnold Schwarzeneggar on the campaign trail, I was struck by his uncanny resemblance to Stan Laurel. Unfortunately, there is no Oliver Hardy to keep him in check.

Regardless, with more than 100 candidates for this election, California politics has become a crazy circus consisting entirely of inept clowns.

It is hard to decide between the candidates of the SCAM party (Sinister Clowns Abort Morals) for Republicans, or the ETG party (Every Thing Goes) for Democrats.

Let us overwhelmingly say no to this idiotic Issa-ism.

Shoab Kamran
Fremont

. . . .

Dean is the people's choice

LATELY, a concerted effort is being made not only by the Republican king-makers, but also the Democratic Party top-dogs, to discredit Gov. Howard Dean any way possible.

I consider him the people's choice to lead the nation.

To judge any candidacy, we need only to ask the following questions:

▸ Does the candidate convey a feeling of sincerity, honesty and loyalty to the country and her people? The plans and the public statements can change with the changing facts and circumstances, but sincerity cannot.

▸ Does the candidate have foresight and vision to chart out methodologies and plans, to provide the three essentials for a modern, successful and secular society, i.e., health, education and jobs for all, regardless of race, religion or gender?

▸ Would the candidate strive to make the country a moral, military and economic strength internationally, such that no country or radical group in the world would dare despise or harm this country or her people?

▸ Does the candidate lead an ethical, moral and honest personal life that not only his/her own children, but all the young people in this country would be inspired by it?

▸ Does the candidate have a proven and results-oriented record of leadership?

Dean has done a commendable job as the governor of Vermont.

He is the people's choice, not only for the Democratic Party leadership, but also for the presidency of this country.

Let us overwhelmingly elect him the next president of the USA.

Shoab Kamran
Fremont

. . . .

Let families rule

LATELY, THE death penalty discussion has been heating up because of the recent two cases in California (Tookie Williams and Michael Morales) where the final decision has been left to the governor of California.

The state government is there to uphold justice for all her citizens in case of wrongs being done to them.

Where the crime is grotesque and evil against other members of the society, and the perpetrator is proven guilty beyond doubt, the justice has to be swift and strict.

In Muslim jurisprudence, after such a criminal is proven guilty, the punishment is left for the wronged party to decide. The wronged party can decide for 'life for life' compensation, or forgiveness.

Of course, a criminal of that nature would be incarcerated and kept away from the society, even if forgiven.

This way a real justice is performed so that the governor or the state's citizens do not have to bear the burden of such decision or guilt.

My suggestion is that we should incorporate the punishment decision by the wronged party as part of our justice system.

Shoab Kamran
Fremont

. . . .

LETTERS TO THE EDITOR

Like Genghis Khan

SEEMS A universal ruler from the West is set on the same strategy and purpose as Genghis Khan did coming from the East about 800 years ago. I guess history repeats itself.

And yet, I am saddened by the loss of life from this war, the overwhelming human misery it will create, and the further devastation to the cultural heritage of that ancient land it may produce.

President Bush may as well repeat Khan's words to Iraqis: "I am the punishment of Allah. If you had not committed great sins, Allah would not have sent punishment like me upon you."

I guess President Bush is banking on the same scenario as the Mongol conquering of Baghdad by seeking revolt from the Christians and the Shi'a population of Iraq.

Here is a paragraph from the historical description of Mongols' conquest of Baghdad in 1258 written by Frank E. Smith in his book "The 6th to 19th Century — Conflict Attitudes and Changing Religion." How ominous and sad.

"In 1258 a Mongol army led by Mongke's brother, Hulegu, attacked Baghdad. Some Christians in Baghdad used the coming of the Mongols as an opportunity to free themselves from Muslim rule or to avenge past wrongs, and Mongols military leaders were willing to use the conflict between Christians and Muslim rulers to their advantage. Baghdad resisted the Mongol advance and was utterly destroyed, with Christians and Shi'a Muslims in the Mongol army reported to have been the most fervent participants in butchering religious opponents — Baghdad's Sunni Muslim inhabitants — with Christians and Shi'a Muslims being spared."

Shoab Kamran
Fremont

Opinion

The Argus

theargusonline.com

FRIDAY
February 28, 2003

Nothing's changed

THE WORLD is again at the brink of war, and 12 years have passed since you published my letter of Jan. 24, 1991.

You may as well print that letter again, as nothing has changed in the obstinate and violent ways of world leaders trying to achieve their short-term goals.

While millions of people have been maimed and many children have been born deformed, died prematurely and are dying of malnourishment daily because of the last war, the warmongers' rhetoric continues.

Are all countries supposed to have equal rights to follow the cultural, economic and traditional values of their people? Are the militarily most powerful countries also the decision-makers as to what are the best human values and the best forms of government for the rest of the world?

Are the violence and wars to continue as a means to satiate the greed for the natural resources?

It is incumbent on the powerful, well-off and decent people to distance themselves from their governments' territorial and economic greed and help bring about a fair and decent solution for the desperate people seeking their rights and liberties.

Using only violent means to avenge and eradicate terrorism or the despotic rulers will never eliminate them.

Only when all communities have a chance to live with a decent measure of civil liberties, freedom and equality will there be a start in getting this world free of tragic terrorist acts.

Only when the leaders of the world understand that their people can hand them over to the World Court for crimes will there be a start to the end of ruthless and despotic rulers.

Shoab Kamran
Fremont

WEDNESDAY September 26, 2001

Stopping terrorism

I AM surprised at Gaurang Desai's Sept. 20 letter.

It seems to trivialize the efforts of the United States and the world coalition to eliminate the terrorism from the world.

After eliminating the immediate known and proven sources of terrorism without spilling blood of innocent people from any country, we should be looking at a long-term view of creating a peaceful future for the people within as well as outside the United States.

The desperate people whose existence is so miserable and hopeless that they have nothing left to lose would in most cases commit atrocities without any regard for their own or anyone else's life. In fact, in many cases they would try to finish themselves in a dramatic fashion.

It is incumbent on the powerful, well-off and decent people of this world to distance themselves from territorial and economic greed, and help bring about a fair and decent solution for the desperate people in the world seeking their rights and liberties.

Using only violent means to avenge and eradicate the terrorism will never eliminate it.

Only when all communities in the world have a chance to live with certain civil liberties, freedom and equality, will there be a chance of getting this world free of tragic terrorist acts.

Shoab Kamran
Fremont

MONDAY
May 12, 2003

Methodical subjugation

MOST PEACE- and justice-loving readers will disagree with the remarks made in a My Word in the April 25 Argus.

Except for the first three and the last paragraph, the column is conjecture to confuse readers about the Iraq invasion. As such, India is also seeking sanctions to attack its neighbors on similar pretexts.

Current Indian government policy is staunchly pro-Hindu rather than secular. A process of subjugating the Muslims and Christians to the lowest Hindu caste is being followed methodically.

The U.S. Constitution guarantees equality and justice for all within the country or without. As it is practiced today, the U.S. foreign policy is to attain short-term goals at any price. For this purpose we place and/or support cunning leaders who soon become monsters, and then we have to destroy them.

Currently, Indian and Israeli rulers have the same U.S. backing to annihilate the freedom-seeking desperate minorities in the name of eliminating terrorism, while we look the other way.

Independent U.N. observers must be placed in these trouble spots to ascertain the realities there. These ruthless leaders must be persuaded to facilitate negotiations that would result in decent solutions for the desperate people seeking self-rule.

The example of East Timor's creation as a country, through referendum, should be followed everywhere. All people seeking self-determination must be given a chance to vote for it, under the U.N. supervision.

Let there be 2,000 countries in the world, if that is what people opt for. And, then, create a harmonious economic relationship among all nations, rather than build controlling hegemonies based on paranoid economic security reasons.

Shoab Kamran
Fremont

The Argus

P. Scott McKibben
President and Publisher

Nancy Conway Vice President, Executive Editor	**Michael Lynch** Senior Vice President/Advertising	**Patrick Brown** Senior Vice President/CFO
Steve Waterhouse Editor	**Dennis Miller** Senior Vice President/Production	**Jim Dove** Vice President/Circulation
Mario Dianda Deputy Executive Editor	**Robert Jendusa** Vice President/Human Resources	**Chris Forsyth** ANG Editorial Page Director

SUNDAY
January 17, 1999
LOCAL-8

LETTERS TO THE EDITOR

Clinton's arrogance and contemp

The Argus

MONDAY, MARCH 29, 1993 ❦ PAGE A-14

Give Clinton
a chance

Editor: I am surprised at your many columnists and readers who have and are continously giving their cynical and prejudicial opinions on the handling by President Clinton of the serious problems facing this country.

Even though I did not vote for Bill Clinton, I like millions of other rational Americans, am willing to give the president time to do something with the mess this country was left with.

The immediate attention, determination and concerns shown by President Clinton to solve the problems of common Americans, have given me and most common Americans the confidence that the American dream may be possible for them and their children, and not just for the priviledged few.

We are willing to sacrifice a little now for the greater benefit of this country and her future generations.

I thank God that the era of the trickle-down greed (economics?) has ended, and that the era of trickle-down morality (justice!) is dawning.

The true and lasting leadership, nationally or internationally, is achieved through practicing justice, equality and compassion for all, and not through selective use of power, greed and callousness.

Shoab Kamran
Fremont

IT IS such a shame that we as a nation have to suffer through this painful process of President Clinton's trial in the U.S. Senate. We have become a laughing stock of the world.

We have our biggest role model for our children not only acting morally inferior to a normal street person, but on top of it, he has been continuously lying about it with extreme arrogance. The example we can give to children now is how not to be like President Clinton.

To err is human, but to keep on committing the same deeds for which you are already being sued is either extreme stupidity (not a chance for such a smart man), or it shows extreme arrogance and contempt for the intelligence and sensitivity of millions of people in this country.

Can a lie be justified when it suits your purpose, or if the truth may have dire personal or public consequences?

Condoning President Clinton's moral character and justifying his lying convinces any decent person that all such fans of Clinton must have the same moral attitudes about life in general, and must have the same commitment toward the laws of the country.

The way President Clinton has shamed his family and country, a Japanese politician, in the same situation, would have committed hara-kiri long ago. The least President Clinton can do is to resign from the presidency and save the nation from further agony, embarrassment and suffering.

As you go up the ladder of public life, so too must the moral and ethical character of the person become superior, so as to be a role model for the living and the coming generations.

A devoted and upright — but not so brilliant — president is far more preferable to a genius who is a manipulator and a self-serving liar.

Shoab Kamran
Fremont

The Argus

P. Scott McKibben
President and Publisher

Nancy Conway	Michael Lynch	Patrick Brown
Vice President, Executive Editor	Senior Vice President/Advertising	Senior Vice President/CFO
Steve Waterhouse	Dennis Miller	Jim Dove
Editor	Senior Vice President/Production	Senior Vice President/Circulation
Mario Dianda	Robert Jendusa	Tom Tuttle
Deputy Executive Editor	Vice President/Human Resources	ANG Editorial Page Director

WEDNESDAY October 27, 1999

MY WORD

School boundary decision must be reversed

I am writing in response to the article headline: "New boundaries drawn by district," in the Oct. 23 Argus.

My family is one of the families affected by the school attendance boundary realignment recommendation.

I moved my family to Fremont from Union City in 1988 to provide a better educational environment for our oldest son, who was to start high school that year.

We chose Mission San Jose school attendance area because of the combination of a higher level of educational environment, peaceful neighborhood and the drug- and gang-free schools.

Our oldest son graduated from Mission San Jose in 1992, and it was a mutually beneficial four years for the school and for him.

Of course, a successful school name breeds further success by attracting the students of the ambitious, industrious and professional parents who believe in pushing their children to the utmost in academic achievements.

They exercised the freedom of choice for the education of their children by selecting to live in Mission San Jose High School area, as long as they were willing to pay a higher price for the location of their home.

Many like my family moved into this neighborhood for only this purpose. The first few years we were so financially strapped that many of our personal needs and family entertainment plans

SHOAB KAMRAN

had to be curtailed to cover the mortgage and property tax payments.

The question is whether the freedom of choice of providing the education parents want for their children has been taken away from the parents who chose to live in a certain neighborhood to provide a good educational environment for their children.

I can definitely say yes, it has

The question is whether the freedom of choice of providing the education parents want for their children has been taken away from the parents who chose to live in a certain neighborhood to provide a good educational environment for their children.

been taken away by the Fremont school district decision to realign the high school attendance boundaries.

Either this freedom needs to be restored by reversing the decision, or a proper compensation for moving costs ought to be provided by the district to those parents who wish to transfer their family to the new Mission San Jose attendance area.

If the affected families' suggestions are ignored, then we collectively have to stand up and be counted by voting overwhelmingly to pass the educa-

tional voucher system in the next election.

This way we would all be able to send our children to whichever schools we want.

And, it will get us back the freedom and equality of choice that we seek — and curtail — or, better still, eliminate — this bureaucratic educational decision-making system.

I do not care if my youngest son, who is in the eighth grade in Hopkins Junior High, goes to Mission San Jose or Irvington High. I know he will do well anywhere.

But my son's mind and heart is set on going to Mission San Jose High, especially since his older brother graduated from there, and most of his friends from Weibel Elementary days and Hopkins Junior will be going to Mission.

When I showed him the news about the decision on boundary alignment, he was greatly upset.

I tried to convince him by giving positive reasons about being in Irvington, but he did not seem to accept it.

It is upsetting for me to see my son so upset.

Maybe the time has come for the educational voucher system to pass, and to establish good private educational institutions in this area, to provide parents an alternative to the mess the Fremont school district is creating.

Shoab Kamran moved to the Mission San Jose area of Fremont in 1988.

Let justice prevail

I HAVE been following the ongoing struggles of parents of mentally disabled children of Serra Residential Center in Fremont from the mid-1990s through The Argus.

As a father of an autistic son, I can empathize with these parents, who are in their 80s.

At this age and stage in the parents' life, it was a consolation for them that justice had prevailed.

The wrong that has been inflicted on them by the National Benevolent Association of the Christian Church cannot be compensated with money. My heart cries out for them.

These parents again are seeking space to establish another facility similar to the one they had provided for their children 25 years ago. They have only one wish: to know that their disabled children, in spite of their disabilities, will spend the rest of their lives in a healthy, happy environment that is mentally, socially and physically nourishing.

They are seeking only about seven acres compared to the 43-acre site they had before.

It was because of the failure of the Disciples of Christ Church, the community and the city of Fremont that such a tragedy took place. The greatness of any society is determined by how it treats its unfortunate, poor and disabled populace.

I am sure the Disciples of Christ Church or the city have seven to 10 acres available to allocate for this humanitarian cause.

It can be sold to the parents or given on a 99-year lease for a reasonable return.

Then the city could lay claim to being one of the best cities in the country, being a fair city not only for the financially well off, but also for those whom money has no meaning.

Shoab Kamran
Fremont

LETTERS TO THE EDITOR

Serra Center award was good new

WHAT HEARTENING news, "Court upholds $8.7 million Serra Center award," in The Argus on Feb. 12.

When Serra Center was in the news about its land being sold for profit, and all its disabled adult children being moved to different locations, I wrote a letter asking for full support from the community for the unfortunate parents of these children.

As the father of a disabled child myself, I could fully commiserate with the plight of the parents, at their advanced ages, having to face the settlement of the future of their disabled children again.

Pains and loneliness felt by the parents of a disabled child can only be understood by those who pass through a similar experience. The mother of one of the residents, after reading my letter in The Argus, even took the trouble to call and thank me for supporting their case.

I never heard of the court case again until, to my surprise, this news story appeared.

With so much condoning of corrupt and immoral actions in the business and political circles these days, I had assumed that the case must have been lost by the parents — not because they were not in the right, but because

they may not have been able to muster enough clout to face the church-backed organization and the rich land developers.

I commend Joseph Alioto, attorney for the parents, for taking this case, and Judge Michael Ballachey for dispensing justice and denying the appeal of the defendants.

The greatness of a society is judged in history, in the long run, not by the riches of its citizens but by how it treats its unfortunate, poor and disabled populace. May God bless Judge Michael Ballachey and provide him bigger and better opportunities for making correct decisions for the citizens of this great country.

Shoab Kamran
Fremont

The Argus

P. Scott McKibben
President and Publisher

Nancy Conway	Michael Lynch	Patrick Brown
Vice President, Executive Editor	Executive Vice President Advertising/Marketing	Executive Vice President Administration
Steve Waterhouse	Dennis Miller	Jim Dove
Editor	Senior Vice President/Production	Senior Vice President/Circul
Mario Dianda	Robert Jendusa	Tom Tuttle
Deputy Executive Editor	Senior Vice President/Human Resources	ANG Editorial Page Direc

SUNDAY October 8, 2000

LETTERS TO THE EDITOR
Get that BART station built

Editor: As a resident of the Tri-Cities area for the past 15 years, I am appalled at the Fremont City Council actions and decisions. Seems that once elected, these officials start working for "special-interest groups."

For the last three years, the Fremont City Council has been trying their best to stop the progress of BART construction on one pretext or another.

Mass transit systems will be the only economic lifeline for the San Francisco Bay Area in the next 20 years. Every day the construction of the BART extension is delayed, we are tightening the noose a little harder on the economic future of this area.

The extreme urgent need of a BART extension to Warm Springs far outweighs the perceived (by whom?) aesthetic and natural blight that BART over Lake Elizabeth will create. In actuality, I would like to ride on the BART train to be able to look at and appreciate the beauty of the park.

The natural beauty of the park will not be lessened any more than all the artificial fences, roads, car parks, pavements and other commercial setups on the park. There is serenity, far-sightedness and beauty in seeing a non-polluting mass transit system getting people and business across to their destinations in a pollution-free, clean, convenient and cheap way.

That is the only way for the San Francisco Bay Area to survive, and the sooner the Bay counties and city governments realize that, the better it is for our future and our childrens' future.

I have yet to see one letter in favor of the Fremont City Council's action against BART, and I strongly urge them to produce people and reasons in favor of their action.

I want to say loud and clear to the Fremont City Council, let the BART project proceed now and immediately. Our patience is running thin.

Shoab Kamran
Fremont

We need BART trains

I CANNOT help but praise state Sen. Liz Figueroa (D-Fremont), for always finding time from her state responsibilities to fight for local causes, as well (extending BART).

She seems to have the courage of her convictions.

In 1992, I had written a letter on the BART extension. At the time, it was the Fremont city government that scuttled the plan by suing BART on extending the line over the Central Park Lake rather than under it. Almost nine years have passed by with no progress on the issue.

The traffic in the meantime has multiplied many times over.

Short-term or special interest measures have to be clearly understood and rejected by the public. Light rail or commuter rail from San Jose to Union City is no solution. These cause additional congestion and frequent stoppages of traffic on the roads.

The most critical area in the

Bay Area is the Interstate 680 corridor.

Adding additional lanes are no solutions, either. By the time they are completed, the traffic will increase to the level that no relief in the flow of traffic appears.

The only solution is an extensive network of BART trains with local feeder transports to the BART stations.

It behooves the state, the county and the city governments to rise above the political interests and prioritize the mass transit projects for the near and long-term future of this area.

Shoab Kamran
Fremont

The Argus

MONDAY, FEBRUARY 1, 1993 ❦ PAGE A-12

India's history needs clarification

Editor: Lately there have been wrong impressions being given on the aftermath of religious violence in India, and distorted historical views presented. The following historical facts need to be clarified here:

1. Muslims ruled India for more than 600 years, and during that time there were fewer incidents of religious persecution than there have been in the last 45 years of so-called secular Indian independence.

2. If Muslim rulers of India were there to spread Islam by force, then India would have had a 100 percent Muslim population. The fact is that Islam was spread in India by religious and pious persons whose monuments are still visited and revered by thousands of all faiths in the Indian sub-continent.

3. The reason that there are fewer people of other faiths in Pakistan is because Pakistan was essentially created from those areas of the Indian subcontinent where there was almost total Muslim population. Unfortunately many other areas in India with majority of Muslims, e.g. Kashmir, were, through political conspiracy, prevented from becoming part of Pakistan. And Muslims in those parts of India are daily suffering worse treatment than the lowest Hindu caste.

All great religions of the world have taught equality of all human beings, regardless of color, creed or religion. I just hope and pray that people of all faiths and ethnic origins throughout the world learn to live peacefully together, and solve their differences without clouding the issues with religious prejudices.

Shoab Kamran
Fremont

Arab nations deserve protection

Editor: Once again, it has been proven that the Arabs are still a medieval tribal society in which millions may follow the hollow rhetoric of a strong and ruthless leader and thousands may be willing to die at his whims.

The Arab nations desire and deserve the protection of the stronger advanced countries. The United States must use firmness for the next few years in helping to cultivate a sophisticated culture and build a political structure for these countries, if they are to be preceived as responsible members of the international community.

Shoab Kamran
Fremont

Two points about Gulf crisis need clarifying

Editor: The events in the last few months have brought the world to war. As my analysis of the crisis, two things need to be clarified for those who are directly involved in the crisis and those who feel obligated to be involved:

✔ The annexation of Kuwait by Iraq is an Arab territorial problem and not an Islamic problem. More than 10 times the Arab followers of Islam are of non-Arab origin, and better Muslims in many instances. If the faith of Islam is not at stake for more than a million Muslims among the 250 million Americans, how can it be in jeopardy with half a million Americans of multiple faiths in Saudi Arabia?

✔ There is no doubt that the United States stands for the highest ideals of any country in the world. But too many times, these high ideals are buried under the debris of covert and military efforts applied to achieve the capitalistic and economic advantage from those countries not strong enough to face up to the U.S.

This is exactly what is happening now: Iraq has a parliamentary government while both Saudi Arabia and Kuwait have/had autocratic rulers; Iraq has freedom of religion and worship while Saudi Arabia does not even allow practice of their religion to those who have come to defend them; Iraq provides freedom to foreign press to interview citizens on the street, while in Saudi Arabia not a single expression has been heard from a Saudi man or woman on the street.

An average Iraqi is better educated than a Saudi or a Kuwaiti. Iraqis have far fewer horror stories on civil rights violations against Third World country workers than the Saudis or Kuwaitis. So it seems that the United States, as so many times before, is fighting to recover and defend the countries that least deserve it and fighting against a country that could be much closer to U.S. ideals than the other two would ever be.

It is time that the United Nations should let the world leaders duel it out with their own lives if the issue is deemed by the majority of the citizens to be not worthy of sacrificing lives. The soldiers of the world should refuse to fight the unnecessary wars brought about by obstinacy, personal egos and greed of the so-called leaders of humanity.

Shoab Kamran
Fremont

www.ingramcontent.com/pod-product-compliance
Lightning Source LLC
Chambersburg PA
CBHW052007280526
45793CB00005B/884